Theory X

The end of the Cretaceous and the extinction of
the dinosaurs

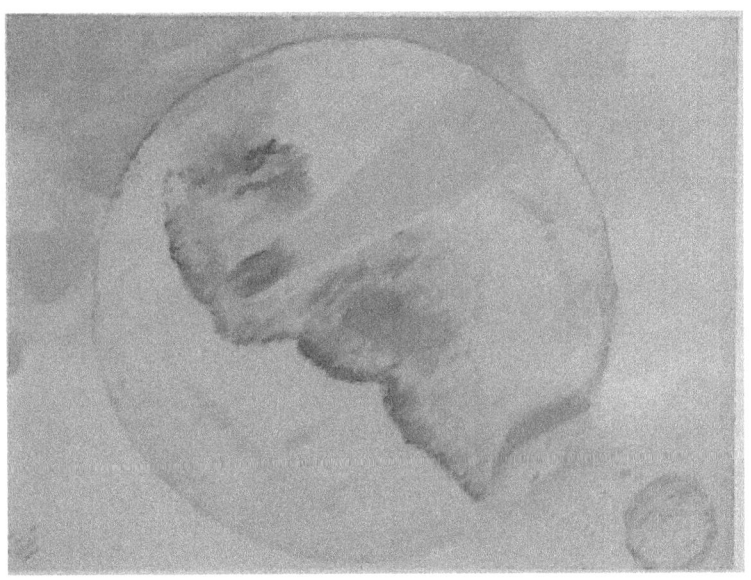

The large impact that disrupted the Earth's crust

CONTENTS

ACKNOWLEDGMENTS

I would like to express my heartfelt gratitude to all those who contributed to the realization of this book. To the readers whose curiosity propels the exploration of unconventional ideas. And to the authors of the 45 hypotheses who shared their knowledge, and from there, further investigations followed, from which I derived the valuable lessons they left me. Your support and engagement make the journey worthwhile. May our collective imagination continue to push the boundaries of what we know.

Theory X

The end of the Cretaceous and the extinction of the dinosaurs

1 PROLOGUE

The content and development of this story refer to our planet Earth and are based on research conducted over several years, with contributions from geology, paleontology, technology; which contributed to knowledge and, of course, to formulate different approaches related to its formation, its phenomena, and its current topography.

The transition from the Mesozoic era to the Tertiary era was surprisingly rapid and enigmatic when considering the multitude of changes and physical and chemical phenomena that occurred, leaving evidence that marked a milestone in the history and evolution of the planet. The episodes leading to the great extinction of life keep many scientists in suspense, trying to uncover the controversial mystery behind the disappearance of the dinosaurs, now fossils; as well as the survival of other organisms, all of which led to the formation of other life systems, with very specific and

unique adaptation processes to new climatic and relief conditions.

This somewhat abrupt transition or change that occurred millions of years ago is today, undoubtedly, the subject of study; therefore, various hypotheses have been proposed that still do not provide accurate explanations about the extinction of the dinosaurs, as a

Theory X

key element and generative topic around which theories arise, and new theories will most likely continue to be formulated. Each of them brings us closer to reality but also moves us away, as objections, new discoveries, and new questions can be found.

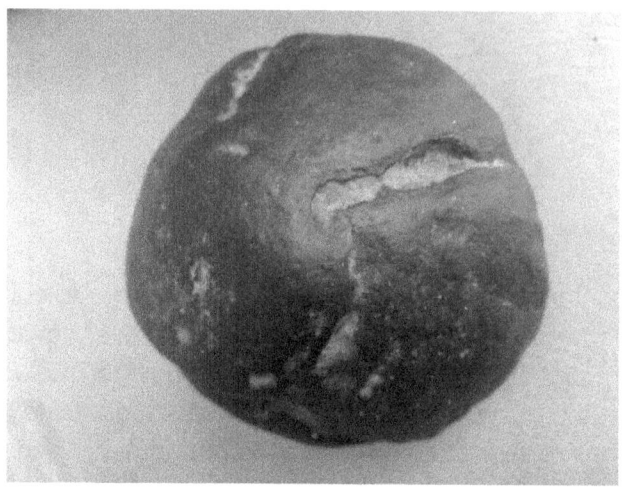

The proposal I make based on the absolute roundness of the Earth, and from its formation until the end of the Cretaceous, it was surrounded by a large mass of water vapor, which, of course, the reader will find in the content and development of this story; will undoubtedly be subjected to rigorous and thorough scientific analysis. I am speaking of the group of scientists, researchers, and scholars of the subject.

After reading my story, analyzing it, and studying it, you, dear reader, will likely agree with my assertions, suggesting that planet Earth, from its formation until the end of the Jurassic, was a planet without mountains, deserts, valleys, plateaus; in other words, its flat surface was designed and better suited for the existence of a monumental population of dinosaurs; with abundant vegetation and an immense population of smaller species, with large lakes, oxygen, and without its lunar satellite; with a rotation perpendicular to the sun and its translational movement in an orbit elongated on one side, the latter caused by the impact that Earth suffered when

Theory X

a meteorite hit exactly where the Gulf of Mexico is located today in the Yucatan Peninsula; pushing it from the inside out of the translational orbit and impacting tangentially from east to west; abruptly slowing down the rotational movement and consequently causing the detachment and sliding of a large part of the Earth's crust, which accumulated in large, medium, small, grouped, scattered, elevated, or low masses, near or far from the impact site. Due to inertia, the vast majority of dinosaurs were thrown into space and then descended, forming large masses of corpses that collected other constituent elements of the planet along their path, such as plankton, water, plants, rock fragments, soil, and minerals, among others. Synergistically, the cooling of the planet and its atmosphere, also due to the impact that pushed it away from the sun, caused the freezing of all that deposited material, which was subsequently covered by the first thaw, initiating chemical and physical processes, beginning the formation of hydrocarbons.

In this way, the planet was organized as we know it today, without ignoring that natural phenomena that change its climatic and topographic conditions constantly appear because we definitely live on a living planet and interact constantly.

The time when these events occurred is still uncertain, but I am confident that sooner rather than later, more evidence will be found, and any of you readers, scientists, will be corroborating the present theory.

It is not irrelevant to respect the opinions surrounding this important topic, wanting to demonstrate once again that the conditions for dinosaurs to survive were not precisely those of our planet today.

Considering, among other things, the many questions that arise about the many and different approaches, we could ask ourselves the following questions: What is continental drift, and what causes it; how did the moon form in the southern Pacific according to scientific theories, why the inclination of the rotational axis and the elongation of its orbit according to Kepler's theory, how were the

Theory X

oceans, islands, mountains, and underwater platforms formed; what is the geoid shape of our planet, what are the abyssal trenches in the ocean floors, how were the deserts formed, why the isochronic pendular movements of oceanic masses, and the most important of all; how did the large number of dinosaurs that existed become oil? Well, dear reader, these and other great mysteries like the Bermuda Triangle phenomenon will be clarified through the proposals that I will present in my theory x. Changing the dynamics of the impact that occurred in the Gulf of Mexico, drawing conclusions and using common sense, logic, and the physical effects resulting from this impact at the end of the Cretaceous, we will provide answers to the questions mentioned above, and why not to the episodes that occurred with Noah's Ark, which, when emerging together with Mount Ararat, east of the Mediterranean, probably never floated, but rather was only designed to protect all the animals gathered there from the climate change that occurred, given its shapes and dimensions; its enormous weight that it had to support and the lack of knowledge of naval techniques made it completely impossible to build it for this purpose.

2 THEORY X

Everything that makes sense around us arises from the very sense that humans attribute to it; giving meaning to life is inherent to humans and arises as a response to questions and the imaginative and creative capacity inherent in the human mind and thought, which in the face of the works created by a supernatural being and the phenomena observed and experienced, allows them to delve, make interpretations, find relationships, debate, corroborate, in other words, to make sense, to find explanations.

We live on this planet orbiting the sun in the here and now, without even imagining the countless phenomena that have occurred since its formation, and perhaps, we have not even considered that at some point in its existence, its rotational spin momentarily paused. I wonder, could this phenomenon have actually happened? Today it is known that indeed there were meteorites that impacted the Earth head-on, producing other consequences that certainly affected or altered the structure and physical conditions of the surface layer.

Logically, this reference point for formulating the theory I propose requires evidence that can be verified from now or in the future.

Let's open a starting point and begin with a day in history, a day when everything could have changed; the beginning of a new era

Theory X

and another that came to an end.

Let's not count how many, but let's imagine that it has been millions of years since that distant day X when our planet was completely spherical without flattening toward the poles as it is today; with a smooth surface, with size and volume greater than it currently presents, probably increasing by one-eightieth, equivalent to the size of the moon because, in the Mesozoic era, it most likely was part of that large segment that detached from the eastern side of

the American continent, more precisely in the place where the Atlantic Ocean is located today, in a vast area spanning a vast area between the American continent and the African continent. I must clarify that on the eastern side of South America, there are large oil reservoirs that were covered by brackish materials after the masses of dinosaur corpses rolled and occupied these places on the planet; Consistent with theories formulated about the formation of the moon, which suggests that at some point it emerged from the southern Pacific.

This detachment of a large part of the Earth's crust that occurred in this place was caused by the Earth's recoil when it slowed its rotational spin, tilted its axis, and moved away from its position in space, leaving behind a land mass that would later become the moon.

Likewise, during the Triassic, Jurassic, and Cretaceous periods of the Mesozoic era, our planet Earth presented from a distance a view with shades of green and blue, corresponding to its vast plains, endless meadows, immense expanses of lush vegetation, large lakes, and some small seas.

Theory X

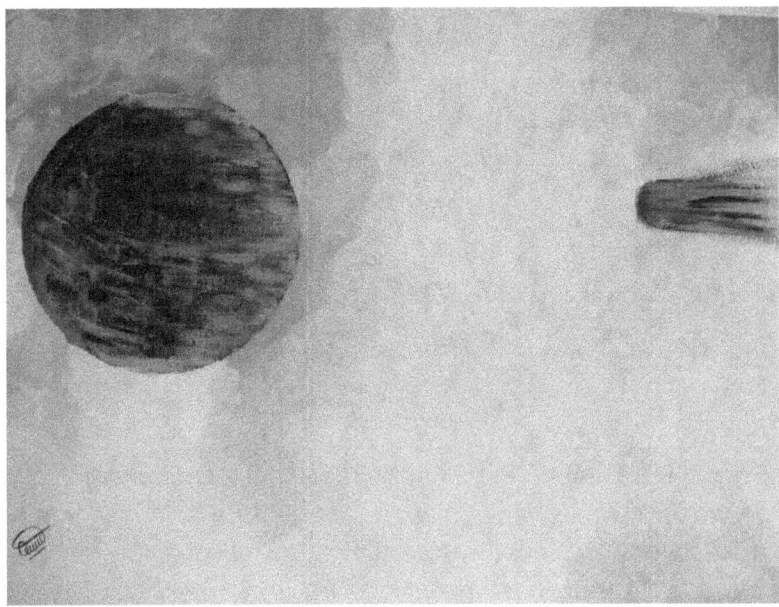

About these landscapes, initially inhabited by the thecodonts, later the dinosaurs settled, populating them as large nomadic herds, respectful of their territories and devouring abundant vegetation.

Their axis of rotation was perpendicular, and their translation in a circular orbit, contrary to the formulations of Kepler's law. The previous characteristics and this last one guaranteed a hot and constant climate throughout the year, very suitable for their vegetation and for the life of the dinosaurs, who received the scorching rays of the sun directly, heating them and penetrating their thick and oily skin, as they very possibly were cold-blooded, able to expose their bodies to long days of heat.

Theory X

The biosphere interacted uniformly with the condensations of a large mass of water vapor that surrounded the planet at that time; a process that occurred during the night, fixing nitrogen on the surface, falling as snow and as permanent dew that allowed for maintaining very deep water levels, meaning that the Earth's crust had, on its entire outer surface, the appearance of moist tropical jungle soil.

There were no clouds, so there were no rains and no rainbow phenomenon; let's remember the covenant of our patriarch Noah, biblical accounts in Genesis, where he specifically and particularly warns his builder to caulk between the joints of the wood to prevent the penetration of cold into this precarious and fragile wooden construction.

On the other hand, he carefully gives instructions for the logical and appropriate dimensions for the construction of a first compartment built on the earth, where all creatures, male and female, smaller in size, were sheltered, excluding the dinosaurs that were not invited by patriarch Noah.

A second compartment or second floor as a storage place for all food reserves, both for the crew and for the animals, and finally a third floor as a living area where it was surely the shelter for all his family, along with the other people who worked on this first major human project to preserve life. We must imagine that Noah's Ark was not that great vessel that the world has always believed in, but on the contrary, an immense house built and anchored on a flat surface, in the middle of a dense forest populated with cypresses native to that place; with solid wood, tight at its joints, probably using vines or wedges of the same wood, likewise using very primitive techniques tailored to the time. Likewise, we must imagine that this giant construction and its designs were not intended to float but to protect the lives of those inside, isolating them from the intense cold that occurred when the planet moved away from the sun. The physical laws of these events could not be violated because it is not true that this precarious and archaic construction could have floated; let's not look at it from the point of view of a miracle, but from the perspective of a physical and real event of the transformation of our planet.

Contrary to our climate system, the planet possessed a vastly different kind of ecosystem; where according to biblical accounts, "water vapor emanated from its interior," assuming that during the night, these emitted vapors condensed, maintaining the water levels of the large lakes and seas that existed and that were an integral part of the outer crust of the planet, characterized by being a moist, fertile surface covered with a large population of gigantic trees in their various and abundant varieties.

The herds of dinosaurs grazed and walked on these types of vast plains, sharing them with other giant predators like the camosaurs, who were hunters par excellence.

There were scavengers and giant birds that flew easily over the planet's surface, perhaps due to a less dense atmosphere.

The inhabitants of the Mesozoic era enjoyed the necessary conditions and an ecological system that provided the dinosaurs, their largest population, with an abundant grazing surface since most of them were herbivores (sauropods, hadrosaurs) and inhabited a wide belt along the Equator, consuming all the

Theory X

vegetation they found along their way and devouring large hectares of plants, which then sprouted in a cycle that occurred every year. They constantly circled the world in a completely nomadic life, doing so independently from each other.

Believing that these immense populations of gigantic pachyderms could have crossed mountain systems or swam across large oceans in search of food would be totally absurd; as well as imagining that they could have lived under the conditions our planet has today.

Dinosaurs were terrible and highly varied reptiles that shared their territories with other giant creatures such as mammals and birds; as well as insects and pests.

According to scientific findings, their reproduction was through eggs, which makes us think that their population was monumental; perhaps there were dominant males responsible for keeping the herds together.

At the end of the Cretaceous, their successful adaptation to different forms of life led to the appearance of other species such as the ceratopsids.

Theory X

The Earth's crust during the Jurassic period of our planet had a larger surface; it didn't have its terrestrial folding; therefore, there were no mountains or mountain ranges, rivers, ravines, plateaus, or valleys; there were also no deserts, seasons, or snow-capped peaks.

Volcanoes probably did exist as they are a natural phenomenon in which incandescent materials are extracted from the planet's interior and moved to the surface due to the accumulation of heat and high temperatures trapped inside the Earth.

The large population of dinosaurs that once inhabited our planet, and which are now part of our consumed petroleum and the petroleum yet to be explored, existed in better survival conditions, adapted to a larger expanse of flat and vegetated land.

Most of them were equipped with a thick and tough pigmented skin layer, deep black in color to dissipate heat. They lived under dense trees that they constantly preyed upon, walked on moist and sometimes muddy surfaces, and their eggs were laid and incubated on large quantities of leaves they cut with their sharp teeth, which fell to the ground forming large nests. Subsequently, all this material deposited on the ground, through decomposition processes, raised the temperature, which, combined with the sunlight that could freely penetrate through the preyed trees, maintained a constant and suitable temperature for incubation.

After the incubation process, mother females took care of and protected their offspring, continuing their journey and advancing through the thick jungles, feeding and dropping branches and leaves for their young to eat. Otherwise, we could assure that the conditions on our planet today are not suitable for a numerous population of large creatures that undoubtedly existed.

Theory X

From there onwards, making an imaginary large-scale calculation, analyzing some of the situations and phenomena that occurred, we could propose the following assumptions:

If we were to move all the mountains of the entire planet where the oceans are today and evaporate their water, uniting all the continents and bringing the moon back to Earth, leaving only plains and large lakes on its surface, we could understand that indeed millions of years ago, dinosaurs could freely graze and survive; to later be buried in the large areas where today there are greater concentrations of hydrocarbons. (This assumption has never been suggested by geologists or researchers in the field).

Using logic, we could think that these planet conditions probably suited their needs better, as on the one hand, there was enough food, water, climatic conditions, easy nomadic movement, and on the other hand, they had the available space to fulfill all the functions of life (respiration, feeding, reproduction, and coexistence).

Another aspect we will analyze is related to the formation of hydrocarbons from the deposits of dinosaur and other species' corpses, as well as the places where there is a greater concentration of them. This leads us to suggest that the phenomenon that occurred as a result of the tangential impact launched them strongly at high speeds into space, only to later fall and randomly deposit in different places, with a greater concentration on the right side of the newly formed mountain systems where the large oil fields are now located. (This assumption has never been proposed by geologists or researchers in the field). The large volumes of these and the way they were buried account for the monumental population that existed; a real explanation for the search for the precious black gold.

So, probably, our planet billions of years ago, its conditions, and respecting the laws of space and the Universe behaved this way; very strange and different from what they are today, but only

Theory X

missing a macro imaginable dimension effect to give it the characteristics that our great blue planet has today.

Thus, we conclude this story on any given day, a day in the life of the Dinosaurs, the end of the Jurassic period of the secondary or Mesozoic era, and the beginning of another.

A day that, due to divine effects, we cannot specify, but let's call it day X. The day with its hours, minutes, and seconds when a great change originated, was the great cataclysm that disorganized the surface of our planet and annihilated the dinosaurs along with other living beings, as they were ejected into space due to the inertia effect that occurred when our planet suddenly, for a moment, slowed its rotational spin, occurring here, "the physical phenomenon that caused the stoppage of our planet's rotational spin" when two colossal forces collided, such as the torque spin that the Earth exerts when rotating on its own axis, against the weight and tremendous speed it imparted by accelerating when impacting the Earth's crust in this way, this giant mass of ferrous materials that is embedded under the mountains of Mexico and not in the depths of the Gulf. With confidence, I dare to assert that in this place where it is located, it constantly projects a magnetization cone towards what is now the Bermuda Triangle, causing phenomena that are not paranormal but the result of magnetic activity, affecting navigation and, consequently, the events that have occurred in this area delimited by three points in the Atlantic Ocean.

Theory X

The Earth, which was completely spherical and smooth, covered in abundant vegetation and inhabited by dinosaurs, is possible, but without being able to assert it, to have formed in such a way that its geological appearance comes from a catastrophic explosion in space. It molded into a massive sphere where the heavier ferrous materials fused in the center as a core, and the lighter parts surrounded it in concentric layers, which later merged under an impact from many comets or masses of water, mixing homogeneously, creating an atmosphere of water vapor that surrounded its environment, giving rise to a planet with ideal conditions for life. Many periods of time passed when the Earth took on a rounded appearance, and along with water, it hardened the finer particles, creating thick layers, many of them coated with fine gold particles.

As an example for this episode that occurred millions of years ago, comparing it with the following experiment: if we were to add a

Theory X

certain amount of water to any amount of sand, soil, or other elements, the particles would join together, and in this way, if we were to scale it up in space, we would say that the Earth formed completely spherical, with a wet layer or crust that, when exposed to the sun's rays, heated up, generating water vapor that condensed during the nighttime. This experimental explanation would give us a macro-imaginable idea, which would closely resemble what the formation of our planet Earth might have been like and which rotated at four times the speed it does today. Suddenly, this rotational spin came to a halt, and its orbital motion around the sun changed its topography, climatic conditions, ecosystem, and caused the extinction of the majority of the dinosaurs.

By way of comparison, let's imagine what could happen if a dump truck carrying sand or another material were to crash and, consequently, abruptly stop when its speed was only 170 kilometers per hour? Logically, an effect would occur, causing its cargo to be pushed towards the driver's cabin, swirling towards the front of the truck's metal cube; some particles would also be ejected through the air forward and to other places, generating one of the laws of inertia, which states that greater weight means greater inertia, and lesser weight means lesser inertia. We could say that something similar happened to our planet, where its entire surface or outer crust literally detached, folded, cracked, and fragmented itself, becoming completely disorganized and chaotic when it received this tangential impact on its surface. This impact was caused by the ferrous core of a small planet that apparently once orbited between Venus and Earth. Since the impact came from the inside to the outside of the solar orbit, elongating it in the opposite direction to the sun's location.

So, if we accept this phenomenon or event without fear of being wrong, we can affirm that its impact force decisively halted the Earth's rotational spin or torque rotation, and due to inertia, everything on its surface was ejected into space at a speed of 4,700

Theory X

kilometers per hour. This means that the gigantic creatures and plants, along with the waters, plankton, and other marine animals from those immense lakes, were ejected with such force and speed that they formed large twisted masses that rolled like huge logs, dragging quantities of material they encountered along the way and swirling in different places on the planet. This resulted in immense cords of corpses and plant debris along the eastern part of the newly formed mountain ranges (hence the most accurate explanation for the location of oil wells on the planet). Other dinosaurs did not share the same fate, getting trapped among the immense trees, some alive, and others as twisted corpses. Many of them correspond to the remains found, which could well have died immediately after the impact or later as a result of the cold due to the significant geological and climatic change that occurred. We should readily accept that this was the beginning of the ice age; the precipitation of a massive mass of water vapor onto the Earth's surface was practically the origin of the oceans, which formed each summer by melting snow, flowing towards the lower parts of the Earth in the form of avalanches, carrying materials where the finer parts ended up forming the shores of the sea.

Theory X

At the same time, the folding of the Earth's crust occurred, and layers of earth up to two or three thousand meters thick detached from the mesosphere, overlapping one another, giving rise to the existing mountain ranges and contributing to the swirling of animals and plants, which originated immense whirlpools rolling easily.

In some dinosaur findings, it has been confirmed that they were found completely twisted in their death position, and analysts presume that their demise was due to chronic constipation. This coincides with the hypotheses presented as number twenty-two and twenty-three, which claim that the dinosaurs suffered a series of digestive disorders that led to their demise.

These hypotheses derived from findings made by archaeologists regarding the strange and twisted position in which their skeletal remains were found strongly support what actually happened. Indeed, the majority of the dinosaurs were ejected into space, turning into large avalanches of twisted corpses. Later, they were

Theory X

buried with sediments carried by the first thawing of the glaciers that had just formed, probably the most devastating, which occurred as a result of the Earth's return to the sun in its equinox of its translational route, causing the snow to melt and undermining the newly formed folding. It was thus demonstrated that because the dinosaurs were immediately buried and completely frozen due to the cold, they did not undergo the process of total decomposition, which, combined with the absence of oxygen, led to the formation of hydrocarbons.

As we can observe, the cooling and heating phenomena that the planet experienced during this period were due to its distance from or proximity to the sun due to the change in orbit, trajectory, or translational route caused by the impact that also diverted it to one side, resulting in greater elongation according to Kepler's theory.

It is very likely that the planet, at a certain point, specifically when it was farthest from the sun at the point of maximum elongation, or properly called its aphelion, and when its folding occurred, leaving it completely disordered, it turned into a huge snowball in space. At the same time, its vapor layer froze and precipitated towards the surface in large snowflakes and stormy rains, thus compensating for its roundness. Later, it would become a planet with abundant water, forming the great oceans, with its two movements: rotation and translational motion, which constantly tilts, causing high and low tides in the sea, maintaining all the required dynamics of being part of the solar system, in the Milky Way.

The eccentricity of the orbit that Earth describes around the sun is defined by science as perihelion for the closest position and aphelion for the farthest position from the sun. It also indicates two important movements that occur in the axis of rotational spin, known as precession and nutation. The latter was discovered by the British astronomer James Bradley.

Similarly, plateaus, continental shelves, and slopes now covered by the oceans were formed each time an avalanche occurred, carrying alluvial material to lower areas. From this phenomenon, we could also say that it contributed to the formation of fine beach sands.

I would dare to affirm that the city of Barranquilla here in Colombia, also known as 'the sandy one,' at some point, when the first thaw occurred, was the mouth of an immense river now called the Magdalena. This phenomenon also occurred in other parts of the world.

It is worth noting that the majority of the Earth's surface became submerged by this enormous liquid mass, giving it absolute roundness, which is typical of any planet in space. Today, science classifies the topography of the ocean floor in the following zones, according to research (Taken from 'The Water and Marine Relief.')

THE CONTINENTAL SHELF: It is the extension of the continents beneath the waters, with depths ranging from zero meters at the coastline to about two hundred meters deep. It occupies approximately ten percent of the oceanic area and is the zone of significant exploitation of oil and fishing resources.

THE SLOPE: It is the area of steep incline that extends from the continental shelf to the ocean floor. It has a depth ranging from two thousand to six thousand meters. This slope, present on all continental shelves, has always experienced landslides and detachments that lead to the formation of large tsunamis.

THE OCEAN FLOOR: It has a depth ranging from two thousand to six thousand meters and occupies approximately 80% of the oceanic area.

THE MID-OCEAN RIDGES: These are elongated uplifts on the ocean floor.

THE ABYSSAL TRENCHES: These are narrow and elongated areas where the ocean floor descends to a depth of ten thousand meters. These enormous fissures likely contribute to the geoid appearance of the planet and formed at the same time as the continents separated, with other smaller ones opening up to become what are now the Mediterranean Sea, the South China Sea, and the Sea of Japan, along with other large seas.

The landslides of land and rocks left the planet completely disordered. A significant part of its outer surface detached from the mesosphere, resulting in the macro effect of Earth's folding, which is the origin of its relief.

Folds have a depressed part known as syncline and the protruding part called anticline, and this leads us to assume the formation of relief. This proposal contradicts the widely accepted but scientifically unproven idea that the origin of relief formation and mountain creation is due to forces operating within the Earth's interior. Another of the significant misconceptions related to mountain formation.

Similarly, as previously explained, with the impact and rock detachments that occurred, the planet shifted and elongated its orbit, leaving behind a vast amount of land in the location of the events. This land corresponds quite clearly to what is now the southern part of the Atlantic Ocean between the two continents, America and Africa. Other fragments of rocks also detached from the southern coast of South America, floating in space. Eventually, they fused with each other, taking on a spherical shape, and began to orbit the Earth, forming the lunar satellite. This large landmass, considered the moon's main nucleus, acquired its gravity, attracting various rock fragments that had remained in its vicinity, shaping its outer relief every time these fragments crashed onto its dusty surface. This was a decisive phenomenon in shaping the lunar exterior, rather than the claim that it was impacted by meteor showers. I want to offer my apologies to all those who, in their research, drew their own conclusions, analyzed, and expressed their opinions, which somehow also became part of my investigations.

Some hypotheses or theories about the moon's formation clearly show many similarities with the characteristics found on planet Earth, such as the isotopic oxygen signature, the presence of silicon and magnesium, mass, and density similar to Earth's mantle. These similarities led some scientists to propose the possibility that the moon emerged from the southern part of the Pacific Ocean at some point. This supports my theory, as I suggest that the large land masses that remained floating in space, or rather at the impact site, once belonged to Earth when it moved away from its initial orbit, leaving behind a portion that detached from the southern part of the planet. Later, it was covered by other rocky materials from the southern Pacific.

Another circumstance demonstrates that the lunar orbit is not aligned with the Equatorial plane, and here lies the possibility of a new explanation for the moon's formation.

In the impact zone where the meteorite collided, an instantaneous resistance force caused the entire Americas, from south to north, to fold up the mantle that had detached from the Asian and Oceanic continents, forming the entire mountain system in the American continent.

Likewise, the collision elongated the Earth's orbit on one side. Also, because it received this impact above the Equator line, the rotation axis tilted by approximately twenty-three degrees, twenty-nine minutes, thirty seconds, as science has indeed measured it, and the Egyptians probably used it as a reference point for building their pyramids.

The tangential impact received above the Equator line allowed the planet to sway and give a slight jerk, causing the Earth's crust to move slightly northward, opening up a large fissure that formed the Mediterranean Sea and the mountain ranges in the Euro-Asian continent. This is where the highest elevation of the mountains,

Theory X

Mount Everest, is located. Possibly, this maximum height was reached when this rocky portion remained suspended in space for a brief moment. Still, gravity eventually pulled it back to Earth, where it became part of the Euro-Asian mountain range. We could say something similar about Noah's Ark if we understood that it was not constructed to float but to withstand the climatic changes that occurred. In the end, it emerged along with the hills of Mount Ararat. Africa separated from America, and a certain part of the land remained at the impact site. When the solar orbit elongated, forming a massive crater on the planet's south, it separated two large continents.

This immense metal mass that struck the planet at the time did so from a northeasterly direction, almost exactly slightly more than ninety degrees west longitude of the Greenwich meridian, above the Equatorial line in northern latitude, over the Tropic of Cancer in what is now the Gulf of Mexico. Searching for it in the depths of the sea will be fruitless, implying that it is located beneath a large number of mountains that formed in this part of the planet.

Some scholars of the subject claim that a meteorite caused a major impact when it hit the Earth, and they assert that this event occurred over sixty million years ago in the Yucatan Peninsula, in the Gulf of Mexico. They attribute it to a large asteroid known as Chicxulub, which collided head-on with the Earth, producing immense thermal energy. From my perspective, what happened in this place was not caused by a meteorite or asteroid but rather by a small planet that orbited between Venus and Earth. By divine intervention or due to the demands of a natural cause, it lost its course or, attracted by Earth's gravity, it headed toward our planet. In its path, it left behind its entire outer layer or cap, retaining only its ferrous core, which tangentially impacted the Earth's surface.

If we change the dynamics of the impact that occurred in this place, we can observe the encounter of two enormous forces that modified the Earth's orbit in its translational movement around the sun, producing a greater elongation on one side, from the inside out. This made the Earth move away a bit more, boldly assuming that it took seven days for the planet to move away once these episodes occurred, at a time when existing climate systems were unknown. According to divine intervention, as announced to the patriarch Noah, 'In seven days, I will make it rain upon the earth,' and considering that at that time, the planet completed its orbit around the sun in three hundred and sixty days. Thus, we assume that since these events occurred, our planet has been gradually approaching the sun again, recovering its initial orbit in two days. This coincides with today's poorly designed Gregorian calendar.

This macro effect that occurred millions of years ago probably destabilized all the elements of the planet, which moved with less resistance, allowing for the easy movement of all the land and rocks that formed the Earth's significant relief. It caused the emergence of tectonic plates that had been under constant and consistent moisture. Thus, we could describe this episode as the origin of continental drift or the separation of continents from the great Pangaea.

As for the greater elongation of the Earth's elliptical or orbital path on one side, according to the astronomer Johannes Kepler, and contrary to what was proposed by Copernicus in the 16th century, in this present proposal, it is considered a valid proposition to assert that the Earth expanded its displacement space and consequently moved a little farther from the Sun at the point of greatest elongation, resulting in lower temperatures. It also affected the way sunlight reaches different parts of the Earth's surface at different times, leading to climatic and seasonal phenomena.

In this line of thought, as the planet and its water vapor atmosphere moved away from the Sun, it cooled and became covered in snow, compensating for the disruption and regaining its roundness. It also regained its rotational and translational movements. When it approached the Sun again in its orbital path, the first glacial thaws and melting of the snow that had covered the planet occurred, eroding the newly formed folds. The entire surface of the Earth was in disorder, with large rocky platforms sliding, folding over one another, and emerging, thus forming all the mountain systems and the new terrestrial relief.

Furthermore, the sediments carried by the thaw formed large avalanches of mud and stone that buried immense rolls of corpses and vegetation, simultaneously forming plateaus and submarine platforms, which over time began to be covered by ocean waters. In these episodes, nature clearly shows us the formation of minerals such as gold from rocks and alluvium.

It is obvious to admit that this event also marked the appearance of rainbows along with rain, and the atmosphere became denser.

Thus, the Earth has been changing over time, separating two eras that left numerous mysteries and some traces of this episode, which marked the end of an era and erased countless pieces of evidence from its origin.

The absolute roundness of our planet is practically determined by the vast oceanic mass that covers it. The islands and continents that emerge from it will, at a later time, possibly be covered by the oceans. This assumes that our planet is getting closer to the Sun every year, trying to regain its initial orbit and, consequently, contributing to global warming.

Continuing with what happened at the moment of impact, assuming that the rotation slowed down and the translational orbit changed, a significant part of the land where Melanesia, Indonesia, Micronesia, Japan, and all of Oceania are located today detached, slid, swirled, and formed the mountain chain in the American continent. The parts that did not detach formed the great archipelago of the Oceanic continent. Larger islands like Japan, Iceland, as well as smaller ones like Hawaii and Bermuda, not to mention all the others, remained anchored in their initial formation sites. The Adriatic Sea

Theory X

is a depression that filled with water. Some seas probably already existed, along with freshwater lakes, which served as watering holes for dinosaurs and other living beings on Earth.

If we analyze further, we can see that the immense island of Madagascar belonged to the southern part of the African continent, along with Australia. The South China Sea and the Sea of Japan were large craters that opened on Earth, very similar to what happened with the Mediterranean Sea and the Red Sea, which were also deeper, more pronounced rifts of greater magnitude.

Similarly, thaws shaped the oceans and eroded the newly formed mountains. With sediments carried from the highest peaks, large valleys, plateaus, and continental platforms were formed. This is also when the formation of hydrocarbons began, accompanied by climate changes like those still occurring today.

The polar ice caps were also formed, gradually thawing over time. The trail of heat and radioactivity that traveled behind this incandescent mass discharged all its power onto the Earth, more precisely onto what is now the Sahara Desert in the northern part of the African continent, extending a wide band of heat along the Tropic of Cancer.

Scientists consider some islands in the Atlantic Ocean near the impact site as volcanic origin islands, which means that they resulted from the heating of the area exposed to the discharge of the millions of megatons that occurred in this impact zone.

The Guajira Peninsula with its large desert in Colombian territory and India, where the heat formed other deserts, give us an idea of the effect caused by the heat and radioactivity that occurred in this zone, above the equator, significantly affecting all these places exposed to radiation.

I must consider that, with the knowledge acquired through science, I have been able to draw conclusions about what those terrible

Theory X

lizards that once populated the Earth were like. The majority of the beings were different varieties of dinosaurs, such as the Sauropods that predominated during the Jurassic period, generally being herbivorous, slow-moving, and traveling the world in search of their food.

Of course, the years for them were not 365 days, but of shorter duration, considering that the solar orbit, being initially completely circular, had a shorter path. Therefore, it is deduced that the climatic system was uniform and without variations in those times.

The voracious lizards, of which we now have vestiges, probably survived that cataclysm only to later perish as their initial habitat became unbalanced. Most of them became extinct by rolling spectacularly on the surface and immediately being buried by the first glacier, the most devastating in history, as it carried all the fractured material from the newly emerged mountains.

Possibly when the Earth moved a bit farther from the Sun, and its elongation increased, it cooled down, turning into an immense snowflake with very low temperatures, where many other species besides dinosaurs died due to the cold. This theory, which has not been dismissed by scientists, aligns with the majority of paleontologists who supported it in the 1950s because their hypotheses suggested a sudden change in the climate system. Others, not as skeptical, thought that a major ice age may have occurred at the end of the Mesozoic era, which could have been fatal for the dinosaurs. However, the hypothesis of dinosaur extinction due to cold was later reevaluated by the Canadian paleontologist L. Rusel, who, along with other researchers, proposed that the rise in sea levels toward the end of the Cretaceous was a determining factor in their extinction.

Today, we have no more than forty-five hypotheses about the possible extinction of dinosaurs, but none of them fully capture the reality. I dare to introduce a new element, which is to consider that, in addition to the cold, the way the dinosaurs were propelled due to

Theory X

the impact and the inertia effect, forming large rolls that mixed with other species and organic and inorganic materials, buried later by those great avalanches, account for and explain in large part the formation of hydrocarbons. This last element is absent in some previous hypotheses but is crucial as a support for this and other hypotheses already mentioned.

In 1950, a book titled 'Worlds in Collision' was published, in which the geologist Immanuel Velikovsky analyzed the deviation of the Earth's axis as a hypothesis based on biblical writings in the book of Joshua, and it is acceptable. Two other hypotheses (number forty-four and forty-five), respectively formulated in 1975 by the geochemist and Nobel laureate Harold Urey and supported in 1979 by Kenneth J Hsu of the Geological Institute of Zurich, proposed the fact related to the impact caused by an asteroid and/or a comet as the cause of the extinction of dinosaurs, algae, and marine plankton.

It is worth noting that these proposals reveal points of agreement with the hypothesis I am presenting, providing support because there are clear or similar similarities to the hypotheses previously proposed.

The world that was completely disordered on that distant day took a long time to recover. The melting ice and rains provided the necessary roundness to keep it spinning, filling all irregular surfaces with water. However, its axis of rotation had become unbalanced, and a greater weight had been added to its outer surface. In other words, all the Earth's crust folded in America on one side of the globe, combined with the weight of this massive iron mass embedded in this location. Thus, the pendulum movement exerted by the vast mass of ocean water, counterbalancing to maintain the uniformity of its rotation, formed the ocean currents that go back and forth in a sudden sway and the tides that rise and fall every twelve and a half hours. This significant natural mechanism that occurs on our planet could most likely be applied in civil engineering, designing a system for tall buildings to maintain a perfect balance oscillation. This situation, as it happens, corroborates the principle proposed by Galileo Galilei, which states, 'the movement of the ocean mass is equal to the isochronism of the pendulum.'

We could also say here that this massive iron structure, which rests well-embedded in the Gulf of Mexico under the fold of its mountains, probably has a lot to do with the phenomenon of tectonics and diastrophism, originating a geological fault there, caused by the weight of this giant and the centrifugal force of the planet's rotation. Why not also provide a relative answer to the great mystery of the Bermuda Triangle, considering the likely reason for this phenomenon, as it most likely originated when this mass of metal cooled down and acquired a certain magnetization that attracts metallic objects that come into its attraction focus.

A good question to consider within this same hypothesis is: Why are the sea levels not the same? Let's go back to the experiment with the sand truck but under different conditions this time, filled with water and set in motion at a constant speed. Of course, when it is set in motion and maintained at a constant speed, due to the inertia effect, the water will move backward, raising the level at the rear of the truck and lowering it at the front. In the same way, let's compare this experience with what happens on the planet. Due to its rotational motion at a known speed today, the eastern part of the American continent, as the rear part of the truck, will receive a larger amount of water from the ocean and will move toward the coasts. Meanwhile, in the west, compared to the front part of the truck, the water levels would decrease, assuming that the American continent is the great dam that separates the two oceans. Also, let's remember that the continents were formed by the impact of this meteorite, approximately seven thousand meters in diameter, which lost its orbit in space and headed toward our planet at hypersonic, supersonic speeds (according to technical language). In its trajectory through space, it gradually merged, leaving a large mass of incandescent metal, followed by other fragments that also impacted other areas of the planet. Its trail of heat, with a devastating effect due to high temperatures, incinerated vegetation, species, and the Earth's crust, forming vast deserts.

Of course, it will be necessary to remove a large amount of sediments that have it buried. Only God and technology will allow us to take a sample of this giant that impacted the Earth. Then, this theory will have been heard.

Another similar or identical phenomenon in its form and consequences would most likely produce a massive tsunami with a devastating effect that would wipe out all creatures from the face of the Earth. The oceans would literally pass over all continents in a violent manner, erasing all forms of life and causing the rotation speed to decrease. Thus, the time would remain the same, but the days would be longer.

These movements of our planet probably gave rise to life, transitioning from inorganic to organic, fulfilling the principle of Oparin.

With this story, I want to provide an explanation for a series of questions that I have not wanted to manipulate. With due respect to the international scientific community, my desire is to subject it to the study and consideration of geological entities, science, and research organizations.

It is needless to clarify that the studies and investigations carried out are the result of more than twenty years of observations and fundamentals acquired as a self-taught individual in the field of geology. The collection of data and scientific information has given me the opportunity to share this macro idea with humanity, also seeking to submit it to the scrutiny of the scientific community, while preserving, above all, the prudence and respect they deserve.

As a result of all these considerations, the scientific community will

Theory X

be forced to change its entire conception of previously studied episodes, starting over and basing it on a possible reality that aligns exactly with the biblical accounts of the Universal Flood. We might indeed have an explanation for the mysteries left by our patriarch Noah and his ark, which was not designed so much to float but to protect all the creatures inside it from the impending harsh weather. Accepting as true that it also appears to have been anchored to emerge along with Mount Ararat, considerations not taken into account because science and religion do not go hand in hand in these events, awakening individual interests in free reasoning.

Most likely, and without getting into controversies, when the Earth rotated four times faster, the duration of day and night would have to be shorter, while when it rotates four times slower, as it does today, days and nights are longer. In conclusion, we would say that, at a higher rotation speed, days are shorter, and humans would live longer, as in the case of Methuselah, and at a lower rotational speed, days would be longer, and humans would live shorter lives. This could explain why these biblical characters reached such chronological ages.

While we know that our planet takes 365 days to orbit the sun, we could propose a new calendar with seven months of thirty days each and the remaining five, the coldest of the year, consisting of thirty-one days, approaching a measure more in line with reality, eliminating leap years, without wanting to discuss our Gregorian calendar.

My work began to take shape in 1995 and was initially just a written form of poetry. It had no support or backing from government entities in my country. I later gained renewed enthusiasm through science and technology, allowing me to exchange my ideas with the intellectual community worldwide. Today, I have the great pleasure of completing this first version, which in some way provides answers to a series of questions that I have been pondering since

Theory X

childhood. Questions like: 'How did mountains, rivers, rains, deserts, continents, oceans, and other phenomena form?' and 'How was the moon formed, and what are its characteristics?' among others. These questions continue to occupy my mind. I hope to continue delving deeper and contributing to science and humanity with new research or the continuation of this macro project. I am somewhat satisfied to find alignments with the proposals of other researchers, making it clear that these were the result of my observations and comparisons in work carried out with study, research, patience, dedication, and continuity for over twenty years. I also conclude that, in a way, as the years pass, the planet has been recovering its circular orbit from the Jurassic era, losing that elongation of eccentricity and getting closer to the sun, trying to regain its initial orbit, and consequently leading to global warming with disastrous consequences, such as the melting of the poles and changing climate systems, resulting in increasingly stormy and hurricane-prone conditions due to the large oceanic evaporation.

This hypothesis about the extinction of dinosaurs, corresponding to number forty-six in its order, emerged from the numerous questions related to our planet, unanswered questions, or hypotheses proposed without any acceptance. It led to the conclusion of the phenomenon that caused the halt in the rotational spin of our planet. This proposal, which encountered many detractors and skeptics, prompted me to create a poem and register it with a notary to establish as true a proposition that was closer to reality, providing answers to a multitude of questions raised up to that point. There were also many warnings and predictions about earthquakes and tsunamis that have occurred worldwide, alongside the volcanic activities that are happening.

Theory X

<table>
<tr>
<td>

Una mañana de aquellas que nacen frescas,

Y en mi cabeza empiezan cosas a pasar,

Se me ocurre que tengo que divulgarlas,

Hoy muy temprano yo se las boy a contar.

Que es lo que pasa dime usted que es lo que pasa,

Si es que este mundo no volverá a girar,

No habrá mañanas que me inspiren a contarles,

Ni noches buenas para un nuevo despertar

Yo me pregunto y es que todos ya lo saben

Que a este universo nunca se le encuentra un fin

Es el misterio que nos tiene navegando,

En un incierto mar de ensueños sin parar.

Aquí en la tierra siempre somos los hermanos

Hijos del diablo no son de Dios,

La guerra es el símbolo que llevamos,

Y no paramos de disparar

Nos olvidamos que tenemos en las manos,

Un mundo amable hecho por Dios.

Yo me pregunto si es que no vale la pena,

Cambiarle al mundo su manera de pensar,

</td>
<td>

Translation:

One morning, one of those that begin fresh,

And in my head, things start to happen,

It occurs to me that I have to share them,

Today, very early, I am going to tell you.

What's happening, tell me, what's happening?

If this world will no longer turn,

There won't be mornings to inspire me to tell you,

Nor good nights for a new awakening.

I wonder, and everyone already knows,

That in this universe, we never find an end,

It's the mystery that keeps us sailing,

In an uncertain sea of endless dreams.

Here on Earth, we are always brothers,

Children of the devil, not of God,

War is the symbol we carry,

And we don't stop shooting.

We forget that we hold in our hands,

A kind world created by God.

I wonder if it's not worth it,

To change the world's way of thinking,

</td>
</tr>
</table>

Theory X

No con canciones ni mensajes que no llegan,	**Translation:**
A la conciencia que no quiere despertar.	Not with songs or messages that don't reach,
	The consciousness that refuses to wake up.
Si lo intentamos y entre todos empezamos,	
Hacer del mundo un paraíso terrenal,	If we try and start together,
Esta es la vida el origen mas humano,	To make the world an earthly paradise,
Obra divina que queremos acabar.	This is life, the most human origin,
	A divine work that we want to complete.
Los que estamos conformando este planeta,	
Los invito a reflexionar,	Those of us shaping this planet,
Guardemos todos los preceptos que nos rigen,	I invite you to reflect,
Leyes divinas sin quebrantar	Let's keep all the principles that govern us,
Así veremos que el mundo que está girando,	Divine laws without breaking.
Por nuestra culpa no va a parar	Then we'll see that the world that's spinning,
Hace millones de años, así fue que sucedió,	Won't stop because of us.
Una gran bola de hierro, a este planeta frenó,	Millions of years ago, it happened like this,
Luego vino el gran diluvio, las montañas socavó,	A massive ball of iron slowed down this planet,
Empezó una nueva vida, todo de nuevo surgió.	Then came the great flood, it carved
	A new life began, everything rose again.

Recorded at Notary Public Office Number Five in BUCARAMANGA, on February 16, 1995, by Dr. FERNANDO MENDOZA ARDILA, Notary.

After those long years of concluding research and hand in hand with the advances in science and technology, I happily came to the conviction that I could become part of a larger hypothesis, presenting and contributing to science an idea of what could be

Theory X

called uniformitarianism or the transition from the Mesozoic era to the Tertiary era.

Next, with the intention of putting into consideration and facilitating the study and comparison of the research conducted to date regarding the genesis of our planet and everything that exists within it, its rivers, its mountains, its deserts; I present a summary of general data about the Earth and the forty-five hypotheses that have been developed to date. I emphasize that only God has allowed all these events to arrange and leave to man, who appears in the Quaternary era, an entire scenario to enjoy, to grow, to coexist, and to bring into play the intellectual capacity that sustains him, improves him, and understands his phenomena.

3 DATA ABOUT OUR PLANET

EARTH FOLDING: These are folds or undulations of layers of the Earth's crust with a folded appearance, where the depressed part is called SYNCLINE, and the protruding part is called ANTICLINE.

TECTONISM: It is the manifestation of forces that operate within the Earth and are of a violent and slow nature, such as DIASTROPHISM.

SHAPE OF THE EARTH: Generally, all the planets in the solar system have a rounded shape, except for our planet, whose roundness is not absolute but presents a flattening towards the poles and a widening towards the equator, determining a difference of forty-three kilometers between its polar diameter and its equatorial diameter, known as GEOID. The fissures that our planet has, scientifically known as oceanic trenches in the depths of the oceans, contributed to the geoid shape of the Earth.

EARTH MOVEMENTS: The Earth has two important movements:

ROTATION: It occurs on its own axis and takes twenty-four hours, producing a torque force.

TRANSLATION: It occurs around the sun and takes three hundred sixty-five days and five hours.

Theory X

LATITUDE: It is the distance from any point on Earth to the equator and is defined as north or south.

LONGITUDE: It is the distance from any point on Earth to the meridian and is defined as east or west latitude.

VOLUME OF THE EARTH: 1,083,000,000 square kilometers.

EQUATORIAL DIAMETER: 12,765 kilometers

POLAR DIAMETER: 12,713 kilometers

EQUATORIAL CIRCUMFERENCE: 40,076 kilometers

POLAR CIRCUMFERENCE: 40,009 kilometers

KEPLER'S LAWS:

The orbit of each planet describes an ellipse, one of whose foci is occupied by the sun.

Each planet sweeps out equal areas in equal times.

The cube of the distance of each planet from the sun is the square of its orbital period.

4 SUMMARY OF THE FORTY-FIVE PROPOSED HYPOTHESES

Within the hypotheses proposed throughout history by different scientists, archaeologists, paleontologists, geologists, and others, they converge and coincide towards a common end or the possible extinction of dinosaurs, without taking into account the corpses that formed petroleum and only analyze their found remains, as a possible common symptomatology for all, leaving the reasoning of all the hypotheses determined under internal and environmental factors of the dinosaurs without a basis. However, several of them coincide with the hypotheses proposed in my book, Theory X.

HYPOTHESES RELATED TO INTERNAL FACTORS.

The initial hypotheses focused mainly on the characteristics of the dinosaurs themselves, appealing to internal factors. All these proposals are debatable as they were not the only organisms that became extinct, and they do not align with other species.

01- Racial Aging: Dinosaurs became extinct due to a natural aging process that produced extravagant and maladaptive traits; this hypothesis is rebutted because, at the end of the Cretaceous period, dinosaurs exhibited high diversity and were successfully adapted to various forms of life. By the end of the Cretaceous, new dinosaurs like ceratopsids had appeared. Considering that reproduction through eggs multiplies rapidly, it wouldn't matter if they died from old age or predation.

02- Excessive Specialization: Dinosaurs became extinct due to over-specialization, leading to maladaptation; this hypothesis was proposed as a complement to the previous one, where over-specialization would accompany racial senescence, associated by Edwin H. Colbert with a genetic factor. Many dinosaurs had already died of old age.

03- Extreme Gigantism: Dinosaurs disappeared due to endocrine failures that led to excessively large sizes, resulting in bodily disorders such as misaligned or dislocated intervertebral discs and bone malformations; this hypothesis was initially formulated in 1888 by Ludwing Doderlein and disseminated by Stephan Zamenhoff. It was the interest in dinosaurs that prompted scientists to conduct new research.

04- Limited Intelligence: Dinosaurs were unable to compete with more agile and intelligent mammals; however, some dinosaurs, like celurosaurs and deinonychosaurs, had large brains and structures indicating great agility. Some scientists suggested that dinosaurs were automatons, prisoners of rigid, genetically programmed behavior, so mammals with flexible and intelligent behavior displaced them from all ecological niches. Paleontologist Edwin H. Colbert accepted this hypothesis, arguing that dinosaurs were virtually walking automatons. This proposal is not comprehensive and lacks veracity.

05- Overpopulation and Suicide: Overpopulation led dinosaurs to develop psychological factors that led to collective suicide, similar to lemmings or some cetaceans; this hypothesis has many objections, as there is no serious justification. Some paleontologists imagined that enormous overpopulation would have led to the development of endocrine and nervous disorders, as well as problems with eggs, where the shell might have become slightly thinner.

06- Overpopulation and Famine: Overpopulation of large herbivorous dinosaurs could have devastated all vegetation, causing a food shortage; however, this hypothesis would have occurred under circumstances that do not exist on our planet today. Ecosystems have self-regulation mechanisms where any imbalances might have affected local populations, but it doesn't explain the simultaneous extinction of other organisms. Contrary to what has been studied, dinosaurs were biologically controlled by other predators, and many died from other diseases, so overpopulation couldn't have occurred.

07- Superpredation: Since large carnivorous dinosaurs like Tyrannosaurus, Albertosaurus, Deinocheirus, Tarbosaurus, Gigantosaurus, etc., appeared in various parts of the world during the Late Cretaceous with great destructive capacity that could have preyed on large populations, this hypothesis is highly unlikely considering that, in addition to dinosaurs, marine plankton also disappeared.

Theory X

HYPOTHESES RELATED TO ENVIRONMENTAL FACTORS ASSOCIATED WITH REPRODUCTIVE ALTERATIONS.

08- Male Infertility: Temperature increases worldwide at the end of the Cretaceous, combined with the large number of dinosaurs, prevented them from adequately dissipating internal heat, resulting in the destruction of male germ cells and widespread sterilization; proposed by R. B. Cowels in the 1940s, it is objected to as not all dinosaurs were large, and many of them perished due to possible planet cooling. Extrapolating the data obtained to an animal of several tons, it was concluded that a global temperature rise would have prevented large dinosaurs from dissipating excess heat. Currently, it is believed that different types of dinosaurs had various thermoregulatory strategies. Conditions were favorable for dinosaur life.

09- Sex Ratio Imbalance: Dinosaurs had a similar sex determination mechanism, and extreme temperature changes resulted in single-sex individuals being born; proposed by Professor Mark Ferguson of the University of Belfast and Edgar Johannes of Louisiana in the 1980s. It is objected to because it is based on living species, and it doesn't explain why these species didn't face the same problem. The temperature affecting the eggs corresponds to a microenvironment where temperature changes within the same nest can be expected, as observed in living reptiles. This hypothesis was proposed in 1980.

10- Thickening of Eggs: Pathological thickening of dinosaur eggshells could have prevented the hatchlings from breaking out; this hypothesis cannot be generalized, as findings of dinosaur eggs suggest the opposite. Additionally, it doesn't explain the simultaneous extinction of other organisms. Entire Late Cretaceous nests have been found in France, composed of unhatched eggs with thickened shells. However, this finding is not generalizable, as other nests from the same period contained hatched offspring. Most dinosaur egg remains, while found in different layers or strata, did not have thicker shells than normal; on the contrary, they were often thinner.

11- Climate Change and Egg Alterations: Another pathology in eggs, with softer and thinner shells, could have prevented the hatching of new offspring due to the lack of necessary calcium for bone formation; proposed by Professor Raymon Dughi of the Museum of Natural History in Aix-en-Provence, he found that most eggshells found in Lower Provence are stratified between two and seven times. This hypothesis cannot be generalized either and does not explain the simultaneous extinction of other organisms. A group of geochemists and paleontologists studied some dinosaur eggs, finding that most of them had anomalies, were infertile, or contained dead embryos, and many eggs were stratified, demonstrating interruptions in shell formation.

12- Nervous Tension and Egg Alterations: Abundant vegetation led to dinosaur overpopulation, and the numerical excess caused nervous tension among females, leading to the loss of their eggs. This hypothesis was developed by Heinrich K. Erbert of the Institute of Paleontology at the University of Bonn, Germany, who worked with thousands of samples from Aix-en-Provence and Corbieres in the Pyrenees. The oldest eggs had thicker shells, up to 2.5 mm thick, while the more recent ones had shells only 1 mm thick.

13- Volcanism and Egg Alterations: Herbivorous dinosaurs suffered

Theory X

a significant reduction in hatching due to ingesting large amounts of vegetation contaminated with volcanic dust containing selenium, leading to the collapse of the food chain.

HYPOTHESES RELATED TO EXTERNAL FACTORS (Gradual Changes).

14- Glaciation or Cooling of the Climate: Intense cold altered vegetation, the primary food source for herbivorous dinosaurs, leading to habitat loss; this hypothesis has something in common with the one proposed in Theory X, as it also presents a climate change when the Earth, due to impact, moved away from the sun, cooling all the water vapor masses surrounding it. This hypothesis also suggests mountain uplift but without valid arguments.

15- Greenhouse Effect: High atmospheric concentrations of carbon dioxide and methane at the end of the Mesozoic era rapidly warmed the planet, leading to the extinction of dinosaurs.

16- Climate and Decalcification: As the climate became warmer and more humid, available oxygen for oxygen-dependent dinosaurs was depleted, resulting in their extinction; proposed by the Russian scientist V. Elissiev, considering the presence of twisted dinosaur skeletons in Upper Cretaceous strata.

17- Habitat Reduction: Large herds of herbivorous dinosaurs found it increasingly difficult to survive as they competed for smaller and smaller areas for food and isolation, leading to inbreeding; however, this hypothesis is objected to as it does not explain the extinction of marine animals and other aquatic communities.

18- Radiation and Blindness: Due to the gradual warming of the planet and excess ultraviolet radiation, dinosaurs became blind and died as they could not reproduce. The arguments in this hypothesis lack serious foundations and were presented in 1982.

19- Vegetation and Excess Food: New plants may have been deficient in some minerals compared to previous ones, causing herbivorous dinosaurs to die from excessive food consumption. This hypothesis lacks credible arguments, as flowering plants coexisted with dinosaurs for at least 100 million years without causing disruptions. Fossilized remains of vegetation do not indicate a drastic change in flora at the end of the Cretaceous. Herbivorous dinosaurs exhibited a wide range of digestive adaptations. It also does not explain the extinction of other terrestrial and aquatic groups.

20- Vegetation and Excess Oxygen: Towards the end of the Mesozoic, flowering plants emitting more oxygen than conifers, cycads, or ginkgoes appeared. An excess of oxygen production occurred on the planet, and dinosaurs were not adapted. Changes in vegetation and the increase in atmospheric oxygen were gradual processes to which different organisms could have adapted. Flowering plants could not have suddenly been harmful; dinosaurs would have shown disturbances long before their disappearance. The fossilized remains of vegetation do not indicate a drastic change in flora at the end of the Cretaceous. This may be the first hypothesis proposed to explain the extinction of dinosaurs. In 1841, Roberth Owen, the paleontologist who coined the term "dinosaur," proposed the theory that during the Mesozoic era, the atmosphere contained very high levels of carbon dioxide and very low levels of oxygen, similar to reptiles. According to Owen, at the end of the Cretaceous period, the current proportions of both gases favored birds and mammals but caused the extinction of dinosaurs.

21- Tough Vegetation: Since hard, leathery flowering plants appeared towards the end of the Mesozoic, herbivorous dinosaurs, adapted to consuming soft and succulent vegetation, died due to maladaptation. This hypothesis lacks a certain argument and lacks serious foundations; the change in vegetation was gradual, and ecosystems have self-regulation mechanisms that would have allowed the emergence of herbivorous dinosaurs with dentition adapted to chew hard vegetation. It also does not explain the extinction of other terrestrial or aquatic groups.

22- Poisonous Vegetation: Towards the end of the Mesozoic, flowering plants synthesizing psychoactive and poisonous aromatic alcohols appeared, and by consuming them, dinosaurs died of poisoning. This hypothesis lacks credible arguments and lacks serious foundations, but it does show dinosaur remains contorted and twisted to support this hypothesis. It is somewhat related to the possibility that they did not die from poisoning but rather rolled awkwardly, as explained in the episodes narrated in this same book, Theory X.

23- Digestive Alterations: Herbivorous dinosaurs, when consuming a new type of vegetation without digestive properties, became extinct due to chronic constipation. British scientist Anthony Hallan, along with Dr. Fritz Khan, determined that ferns contain laxative oils that could have harmed dinosaurs, causing death from chronic constipation. This hypothesis was poorly formulated, rendering it completely invalid, and it was presented with a humorous tone.

HYPOTHESES RELATED TO BIOLOGICAL CAUSES

24- Predator Caterpillars: During the Cretaceous period, caterpillars must have expanded dramatically, leading to overpopulation that devoured all the plantations, leaving herbivorous dinosaurs without food. This hypothesis lacks its own basis if we understand that caterpillars caused defoliation of all vegetation, and if lepidopterans were so abundant, many of their fossil remains would be found. In reality, their remains are very scarce and only appear much later. Ecosystems have self-regulation mechanisms, and such imbalances could have affected local populations, not all dinosaurs and a diversity of other terrestrial and aquatic organisms. This hypothesis was formulated in 1962.

25- Egg-Eating Mammals: Some fossilized dinosaur eggs appear to have been opened and emptied, and during the Mesozoic, small nocturnal mammals possibly fed on eggs. Dinosaurs went extinct because their eggs were devoured by Morganucodon or other small mammals. Mammals only became dominant after the extinction of dinosaurs. Dinosaurs had coexisted peacefully with mammals for about 160 million years, and there is no reason why they would suddenly start eating eggs, not just any dinosaur's eggs but all types of dinosaurs. It does not explain why other egg-laying reptiles or birds did not disappear, nor does it explain the extinction of marine faunas like ichthyosaurs, which gave birth. Ecosystems have self-regulation mechanisms, so egg-eaters would not have exterminated them; such imbalances could have temporarily affected local populations but not all terrestrial and aquatic organisms. There are currently species that feed on the eggs of others without endangering the species whose eggs they feed on. This hypothesis cannot be analyzed solely based on the findings made during the 1922 expedition to Mongolia, where some eggs were found opened and emptied, leading to the conjecture that they were consumed by small mammals.

26- Parasites and Epidemics: When new land routes for migrations were formed, herds of dinosaurs came into contact with populations, transmitting parasites and microorganisms among them, causing their deaths. Epidemics cause massive but not general deaths and leave evidence, which could not have happened to dinosaurs.

HYPOTHESES RELATED TO GEOLOGICAL CAUSES.

27- Increase in Gravity: At the end of the Cretaceous period, the mass of the Earth increased, and dinosaurs, given their large size, could not withstand the change because it made them too heavy. This proposal has not been put forward by scientists but by speculative popularizers. On the contrary, gravity could not have increased if a large portion, say one-eightieth, broke off to form the moon, which means that the Earth's mass decreased, and consequently, gravity decreased.

28- Reversal of Magnetic Poles: The reversal of magnetic poles would have caused the entry of large amounts of ionizing radiation from the sun, resulting in deaths, increased mutation rates, sterilization of fauna, and possibly fluctuations in gravitational constants and atmospheric composition and pressure. R. Uffen in

Theory X

1963 suggested that changes in terrestrial magnetic polarity would be accompanied by a decrease in the Van Allen belts. The reversal of magnetic poles would have caused the entry of large amounts of ionizing radiation from the sun, resulting in deaths, increased mutation rates, sterilization of fauna, and possibly fluctuations in gravitational constants and atmospheric composition and pressure. It does not explain the survival of other groups or the extinction of marine forms that would not have been affected by the radiations. Pole reversal processes have occurred quite frequently without being associated with mass extinctions. It has been suggested that magnetic pole reversal processes may be related to meteorite impacts; in this case, it would not be the cause but would be related to a different cause. It is very true that until the end of the Cretaceous, the Earth's atmosphere was different; it could have been a vast amount of ice or water vapors that fell on Earth, eroding the entire surface and forming the oceans. So, this proposal is not far-fetched.

29- Destruction of the Ozone Layer: The ozone layer was destroyed by the presence of hydrochloric acid from volcanic gases, resulting in the disappearance of the protective shield against ultraviolet rays. This hypothesis was proposed by M. L. Keith, a geochemistry professor at the University of Pennsylvania, and does not explain the simultaneous and massive extinction of aquatic organism groups.

30- Climatic Changes Due to Continental Drift: Continental drift transformed continental areas into coastal ones and vice versa, with continents moving closer to and farther from the poles, causing major ecological climatic changes that led to the extinction of dinosaurs. The displacement of continental masses separated the larger continents and divided lands into smaller areas, altering ocean currents, wind circulation, and climatic patterns. At the end of the Cretaceous, the sea experienced a significant retreat (regression), abandoning epicontinental basins, which produced more extreme climates and seasons. This episode is more than logical since it happened, but it was only one more companion in the whole process that occurred as a result of this environment.

31- Overflow of the Atlantic Ocean: The extinction of dinosaurs resulted from the overflow of the Arctic Ocean because the colder and lighter Arctic waters, when poured into the warmer and heavier Atlantic waters, formed a layer of cold water over warm and saltier waters, causing a global wave of cold and drought. There would have been a decrease of about ten degrees in the temperature of oceanic water worldwide. This proposal, its author, discarded all internal and external elements accompanying mass extinction, based on the conclusion that dinosaurs died from the cold. But it is evident that a climatic change occurred, and this happens because of the proximity or distance from the sun, which determines the intensity of cold or heat. Since the Earth cools more at aphelion than at perihelion.

32- Orogeny and Reduction of Lakes and Swamps: Changes in the Earth's crust resulted in a reduction in lakes and swamps, leading to the extinction of herbivores inhabiting the planet, such as sauropods and hadrosaurs, followed by the extinction of carnivores. Currently, it is considered that different groups of dinosaurs inhabited the mainland, including large sauropods and hadrosaurs. We cannot speak of mass extinction because dinosaurs never went extinct; they only turned into buried corpses that gave rise to oil.

33- Supposed Emergence of the Moon: It is believed that the Moon emerged from the bed of the Pacific Ocean at the end of the

Theory X

Cretaceous period, causing significant planetary disturbances that led to the extinction of dinosaurs. There is no evidence that the Moon came out of the Earth. In Devonian corals, the effects of tides on daily growth can be seen, demonstrating that the Moon existed 330 million years before the extinction of dinosaurs. It is currently accepted that the Moon originated from a collision between Earth and a Mars-sized body, many millions of years before the age of dinosaurs. The controversial and polemical aspects generated over time regarding the formation of the Moon point to an episode that emerged as the end of the Cretaceous, but what we can reasonably say is that the Moon was like a result or complement that accompanied the other episodes that occurred as a result of the tangential impact of this small ferrous nucleus that was dislodged, thus impacting.

34- Sea Level Drop and Habitat Disappearance: Regression of the oceans seriously affected ecosystems and led to the rapid disappearance of habitats in low coastal plains, drying up and significantly reducing the surface occupied by the seas, reducing the volume of photosynthesizing marine vegetation, increasing the concentration of carbon dioxide in the atmosphere, and making the climate warmer and more variable (greenhouse effect). It does not clearly explain all the patterns of extinction and survival. Some researchers believe that, on the contrary, at the end of the Cretaceous, sea levels rose, which would have caused the extinction. The formation of oceans occurred through the planet's thawing whenever Earth approached the equinoxes; these thaws also carved the Earth and formed rivers, streams, springs, and, in general, water resources that are widespread all over the Earth.

35- Ecological Chaos Due to Geographic Factors: The geographical conditions at the end of the Cretaceous allowed for massive migrations of dinosaurs, which moved from one continent to another, causing ecological chaos by bringing diseases and parasites, competing with other species for food, and facing new predators. It is not possible to think of the displacement of these giants across oceans to reach other continents.

36- Shift in the Earth's Rotational Axis: The shift of the planet's rotational axis could have caused the extinction of dinosaurs and a range of other marine and terrestrial organisms at a planetary level. This hypothesis, proposed by Dr. James Heirtzeler in the 1960s, has much to do with the arguments of the last hypothesis proposed by Jorge Luís Arévalo Durán. Indeed, this event occurred, but it was not the effective trigger, as the phenomenon could have contributed to better climatic conditions and, given its severity, would not have caused massive damage to dinosaurs.

37- Massive Volcanic Eruptions: A massive volcanic eruption in the Deccan region altered the planet's climate and ecology, creating a dense cloud of dust and ash that covered the sun, extinguished plants, and caused intense cold and temperature drop, leading to the extinction of dinosaurs and other groups. Carbon dioxide clouds and emissions of sulfur gases and various acids could have poisoned the atmosphere and rivers, causing the overheating of the Earth's surface. Volcanic dust impregnated with selenium could have caused a significant reduction in the birth rate of herbivorous dinosaurs, followed by the death of all carnivores.

38- Planetary Evolution: Mass extinctions are triggered by endogenous processes caused by the planet's own evolution. At the end of the Cretaceous, there were certain changes that altered the Earth's crust structure, but how could this have influenced the extinction of dinosaurs? It is logical that when the Earth's crust was destabilized, a planetary evolution would occur, a change that we could call an additional episode that contributed to this evolution but was only an accompanying aspect in a participatory event for evolutionary processes.

39- Changes in Trace Element Concentrations: At the end of the Cretaceous, there was an alteration in trace metal element concentrations at the planetary level, affecting dinosaurs. These trace element concentrations date from the end of the Cretaceous, and it is certain that they originated, but they were not determinants in the massive extinction of dinosaurs or other species; they were only companions in the disaster and were part of the events of uniformitarianism.

40- Dust and Interstellar Gas Cloud: When passing through the solar system through a dense cloud of interstellar dust located in the galactic mid-plane, radiation reached Earth, darkening the sky for a certain time, causing death from cold in ectothermic animals and destruction of flora due to the lack of photosynthesis, followed by the destruction of herbivores and consequently carnivores. The dense cloud could have persisted for months or years, causing the death of vegetation, followed by herbivores and carnivores. The better-suited survivors would have been smaller animals. These proposals related to mass extinction could well have occurred, but the other questions that arise and accompanied other episodes such as the formation of the lunar satellite, continental drift, the ice age, in short, nothing matches, and nothing proves that this event occurred in this way. It is very likely and perhaps certain that Earth has been impacted thousands of times without catastrophic consequences, only the stopping of Earth's rotational rotation would generate situations of massive and conclusive destruction capable of causing significant changes on our planet.

Theory X

41- Lunar Volcanism: At the end of the Cretaceous period, lunar volcanic activity sent a rain of these particles to Earth, which, upon entering Earth's atmosphere, blocked the sunlight to a sufficient degree to cause global cooling, which dealt the final blow to dinosaurs. We could say that the moon emerged from the Earth at the same moment they went extinct, given that the moon appeared after the Cretaceous period, and its consequences could not have been so catastrophic as to cause massive disasters and destruction.

42- Supernova Explosion: The explosion of a supernova near the solar system destroyed the ozone layer and allowed the passage of radiation, increasing mutation rates and causing sterility in all large animals, while smaller animals that could have hidden, nocturnal ones like mammals, and those in deep waters would have survived without problems. The turbulence generated affected the atmosphere's heat reduction capacity.

43- Intense Solar Activity: An unusual increase in solar activity could have caused the death or sterility of various terrestrial organisms. Considering that the Earth cooled down, this hypothesis cannot be determined as true.

44- Impact of a Meteorite: A meteorite about ten thousand kilometers in diameter impacted Earth 65 million years ago at a speed of 25 kilometers per second, causing a nuclear winter. First, the incandescent remains fell on forests and grasslands, causing fires that covered seventy percent of the continents, interrupting photosynthesis and reducing the oxygen index to almost zero. Strong winds, torrential rains, hurricanes, and earthquakes followed. Then, a dense cloud formed, composed of a mixture of water vapor, released gases, dust, rocky residues, and metallic elements. Its volatile components, suspended in the air and mixed with smoke produced by the friction of the meteorite with the atmosphere, enveloped the planet in a gigantic impenetrable cloud that spread throughout the atmosphere, preventing the passage of sunlight. This led to a sharp drop in temperature, which fell worldwide to an average of minus nineteen degrees Celsius. Lakes froze, and all species of plants disappeared. The cloud could have persisted for months or years, causing the death of vegetation, followed by herbivores and carnivores. Those most capable of survival were smaller animals. These proposals related to mass extinction could well have occurred, but the other questions that arise and accompanied other episodes such as the formation of the lunar satellite, continental drift, the ice age, in short, nothing matches, and nothing proves that this event occurred in this way.

45- Impact of a Comet: The impact of a comet on Earth caused a massive nuclear explosion, a rapid increase in temperature, and possibly released cyanide, a poison found in the crystalline head of some comets, causing the death of dinosaurs.

46- X Theory: This hypothesis, proposed based on the tangential impact of the ferrous core of a small planet that orbited between Venus and Earth, consequently slowing down the rotational spin of our planet, does not appear in the records of experts in the field. I apologize, hoping that it somehow becomes part of the annals of history as a different and more equitable explanation of reality. This hypothesis includes other proposals that were part of the culmination of a series of joint episodes that radically changed the Earth's ecosystem at the end of the Cretaceous, the era that

Theory X

marked the enigma, erasing traces and triggering mysteries that are still unresolved. In my opinion, considering universal time rather than the time measured by the Earth's orbit and rotation, which can be uncertain, it's impossible to pinpoint an exact date. But there is still much research to be done; we are not even halfway, not even at the beginning of a new era that radically changes what once happened - the end of the Cretaceous, the massive extinction of life that marked the beginning of the Tertiary era under events that erased the past, leaving a point of uncertainty called uniformitarianism where nothing is known about before or after. Generally, we always dare to launch hypotheses, theories, and assumptions proposed by others, but we never reach an agreement because, logically, it couldn't happen. For example, consider the idea of suctioning water from the oceans and sending it out of the Earth; this is impossible. But if it were to happen in reality, what would you think, dear reader? What would you imagine could happen to the Earth if the synchronous movements of the tides didn't occur or if the Earth rotated five times faster than it currently does or if the orbit wasn't elliptical and wasn't tilted? The Earth may have rotated this way at some point, very difficult to prove but not impossible to imagine.

How far are we from thinking about the preservation of life if this world were to end? We hardly have an idea about these events that occur by divine intervention, realizing that supremacy and universal laws do not originate from humanity but from the beyond, from the infinite space that makes us believe we are not alone and that we have barely begun to understand the emergence of life. Only six thousand years of history have passed, and the preservation of the human species occurred with Noah's Ark, protecting not only human life but also that of animals, with the purpose of destroying an entire vital environment to bring about significant life changes. Thus, it is clear that those who were part of these episodes were human beings and could also be part of the future if this world were

Theory X

to be destroyed again.

But let's give a better interpretation to these biblical narratives, and let's imagine a humble Noah dressed in animal hides, a nomadic hunter wandering the world with his companion or spouse, perhaps we would call her Noelia, and a brood of children chosen by the creator. The story says it began over six thousand years ago when the construction of this enormous ark started. Similarly, today, it is being constructed in space as the great space station where a significant portion of the world's population will place their lives while the Earth once again stops its rotation, preventing total destruction as the Earth disintegrates due to the centrifugal force generated by an increase in rotational speed. It's not an unfulfilled prophecy; it's the result and continuity of the physical phenomena that often occur in outer space.

As the intellectual author of Theory X, a product of personal research conducted for over twenty years, I have wanted to disseminate, refute, or approve it to give it validity. If you, dear reader, wish to discuss or approve it, you can write to the following address: 90 Street No. 22-36, Diamante II, or email ardujol@hotmail.com in Bucaramanga Colombia.

ABOUT THE AUTHOR

Jorge Luís de la Santísima Trinidad Arévalo Durán.

Agricultural technologist from the Industrial University of Santander. Musician, composer, writer, poet, singer, and sports enthusiast. Restless thinker about the transformation of gravitational and rotational forces into electrical energy. Author of numerous projects such as:

Directing wastewater to empty oil wells to reduce pollution and methane sources.

Building a helicopter without its rear rotor for greater maneuverability.

Mass cultivation of insects as protein sources for birds.

Transporting waste to rural areas for processing into organic fertilizers.

Legislation project for nature compensation. Since every individual begins to pollute the planet at birth, they should plant a tree or, alternatively, pay for the planting of one.

Legislation project on controlling human procreation. Quality of life should be considered in procreation.

Throughout history, humans have always sought a connection between the largest and smallest things. By finding it, they will have taken a significant step, giving their imagination a breath and looking toward the space that is patiently waiting for them.